S0-BYR-429

DATE DUE

Animal Camouflage

Hiding in a Desert

Patricia Whitehouse

Heinemann Library
Chicago, Illinois

© 2003 Heinemann Library
a division of Reed Elsevier Inc.
Chicago, Illinois

Customer Service 888-454-2279
Visit our website at www.heinemannlibrary.com

Designed by Cherylyn Bredemann
Printed and bound in the United States by Lake Book Manufacturing, Inc.
Photo research by Kathryn Creech

07 06 05 04 03
10 9 8 7 6 5 4 3 2 1

Library of Congress Cataloging-in-Publication Data
Whitehouse, Patricia, 1958-
    Hiding in a desert / Patricia Whitehouse.
        p. cm. -- (Animal camouflage)
    Summary: Describes how animals and insects living in the desert use
various forms of camouflage to survive, capture prey, and avoid predators.
        Includes bibliographical references (p.   ) and index.
        ISBN 1-40340-796-7 (HC), 1-40343-186-8 (Pbk)
        1. Desert animals--Juvenile literature. 2. Camouflage (Biology)--Juvenile literature.
        [1. Desert animals. 2. Camouflage (Biology) 3. Animal defenses.] I. Title.
    QL116 .W45 2003
    591.47'2--dc21

                                                        2002010280

Acknowledgments
The author and publishers are grateful to the following for permission to reproduce copyright material: p. 4 Gordon Whitten/Corbis; p. 5 G. C. Kelley/PhotoResearchers, Inc.; pp. 6, 7, 9, 30T Michael & Patricia Fogden/Corbis; p. 8 Peter Johnson/Corbis; p. 10 Richard Shiell/Animals Animals; p. 11 David M. Schleser/Nature Images, Inc./Photo Researchers, Inc.; pp. 12, 13 Maslowski/Photo Researchers, Inc.; p. 14 C. K. Lorenz/Photo Researchers, Inc.; p. 15 O. Alamany & E.Vicens/ Corbis; p. 16 Alain Dragesco-Joffe/Animals Animals; p. 17 Alain Dragesco-Joffe/Oxford Scientific Films; p. 18 Larry L. Miller/Photo Researchers, Inc.; p. 19 David G. Campbell/Visuals Unlimited; p. 20 Marty Cordano/Oxford Scientific Films; p. 21 Raymond A. Mendez/Animals Animals; p. 22 Gary Meszaros/Visuals Unlimited; p. 23 Rick & Nora Bowers/Visuals Unlimited; pp. 24, 25 Tom McHugh/Photo Researchers, Inc.; pp. 26, 27, 30B David A. Northcott/Corbis; p. 28 E. R. Degginger/Animals Animals; p. 29 Zig Leszczynski/Animals Animals.

Cover photography by Michael & Patricia Fogden/Corbis.

Some words are shown in bold, **like this.** You can find out what they mean by looking in the glossary.

To learn about the Peringuey's adder on the cover, turn to page 6.

Contents

Hiding in a Desert

Many animals live in the desert. Some desert animals are hard to see. They use **camouflage** to help them hide.

Some animals hide so they do not get eaten. Others hide from animals they want to catch and eat. There are many ways to hide. The desert iguana has **disruptive coloration**.

Hiding in the Sand

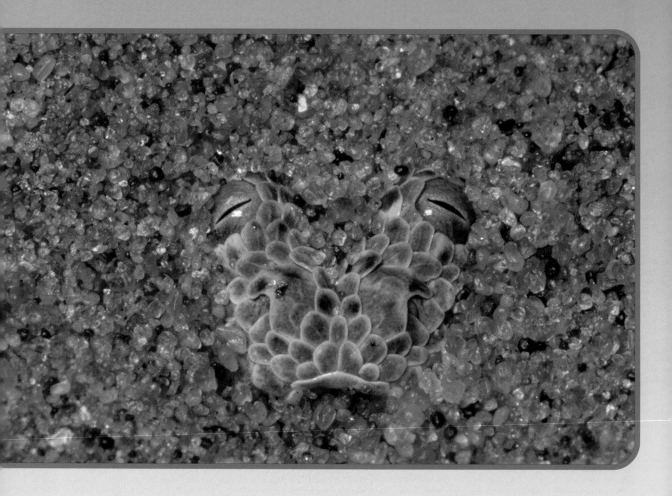

A snake called Peringuey's adder is hiding here. The adder has covered itself with sand. You can see only its eyes. It can surprise the animals it wants to eat.

When the snake comes out of the sand, it is still hard to see. It is the same color as the sand. Animals that look like their **habitat** have **cryptic coloration.**

Hiding on the Ground

Animals that live on the ground must hide from **predators.** Sand grouse lay their eggs on the sandy, rocky ground. It is hard to see a grouse with its eggs.

Sand grouse **chicks** have **cryptic coloration**, too. They look like their **habitat**. Predators cannot find them.

Looking Like the Desert

The sand grasshopper hides on the sand. Its **cryptic coloration** makes it look like sand. **Predators** might think the grasshopper is part of the desert.

The creosote grasshopper has cryptic coloration, too. But it lives in a different part of the desert. Its **camouflage** helps it hide on creosote bushes.

Hiding in Desert Plants

Ash-throated flycatchers live in many kinds of **habitats.** Some of them live in the forest. Other ones live in the desert.

The flycatcher does not need **cryptic coloration**
in the desert. It hides in the saguaro **cactus.**
The flycatcher makes its nest inside the cactus.

Hiding by Looking Different

Some animals have color **patterns** that break up their shape. This is called **disruptive coloration.** This quail has stripes that make it hard to see the whole bird.

This perentie has disruptive coloration. The spots
on its body break up its shape. The perentie looks
like the rocks and sand. The perentie hides so it
can catch **prey**.

Hiding to Hunt

The fennec fox uses **camouflage** to hunt at night. Its light-colored fur looks like sand in the moonlight.

This sand cat was hiding on the sand.
Its brown fur helped it hide. The snake did
not see the sand cat until it was too late.

Looking Like Sand

The sand in this desert is white. The earless lizard is the same color as the sand. This helps it hide from **predators.**

Most pocket mice are tan or gray. Their fur looks like the sand in their **habitat.** These **camouflage** colors help pocket mice hide.

Pretending to Be Another Animal

When stink bugs see a **predator,** they stand on their heads. Then they use a smelly spray. This makes predators stay away.

This beetle does not have a smelly spray, but it **mimics** the stink bug. It even stands on its head. A predator will stay away from the beetle, too.

Hiding During the Day

This is a bird called a poorwill. Poorwills
are most active at night when it is dark.
It is hard to see anything at night.

But during the day, the poorwill needs to hide. Its **cryptic coloration** helps it hide among the grass and rocks.

Hiding by Staying Still

This jackrabbit's fur looks like the ground in the desert. The jackrabbit can stand still for a long time. **Predators** might not see it.

The sand gazelle lives in the desert. Its tan fur is the color of the desert sand. When the gazelle stands still, predators have a hard time seeing it.

Changing Colors

Horned lizards can hide in different parts of the desert. That is because they change colors. This horned lizard looks like the brown rocks around it.

The horned lizard has changed colors. Now
it can hide on this gray rock. It has **cryptic
coloration** everywhere it goes.

Surprise!

The tan scales of the armadillo lizard help it hide in the desert. But it has another way to **protect** itself from **predators**.

To protect its soft belly, the armadillo lizard curls up into a ball. Its spiny scales are on the outside. It is hard for a predator to eat a curled-up armadillo lizard!

Who Is Hiding Here?

What animals are hiding here?
What kind of **camouflage** do they have?

For the answer, turn to page 9.

For the answer, turn to page 27.

Glossary

cactus   desert plant with spines. Spines look and feel like pins.

camouflage   use of color, shape, or pattern to hide

chick   young bird

cryptic coloration   colors that make an animal look like the place where it lives

disruptive coloration   pattern of colors on an animal that makes it hard to see the whole animal

habitat   place where an animal or plant lives

mimic   one animal looks and acts like a plant or another kind of animal

pattern   colors arranged in shapes

predator   animal that eats other animals

prey   animals that are eaten by other animals

protect   keep safe

More Books to Read

Arnosky, Jim. *I See Animals Hiding.* New York: Scholastic, Incorporated, 2000.

Galko, Francine. *Desert Animals.* Chicago: Heinemann Library, 2002.

Kalman, Bobbie. *What Are Camouflage and Mimicry?* New York: Crabtree Publishing Company, 2001.

Index